Jarrold Aquatic Life Series Book 2

Text and Photographs by
Heather Angel, MSc, FIIP, FRPS

Life in our Estuaries

Jarrold Colour Publications, Norwich

1. KNOTS (*Calidrus canutus*) in the Cheshire Dee Estuary. At high tide, waders, such as knots, are pushed high up the shore; but once the tidal level begins to drop, they move down to feed on the exposed mud-flats. Knots are winter visitors which move north to breed in the Arctic. They feed on worms, crabs, and small molluscs as their short legs prevent them from wading out into the water to feed. The sight of a closely packed flock of knots wheeling over an estuary on a bleak winter's day, appearing first dark and then pale as they reveal their undersides, is an unforgettable sight.

An estuary is a place where fresh water flowing down the river mixes with the salt water flowing up from the sea. Here, the river, the land, and the sea come together to produce a coastal habitat which changes not only with the seasons, but also with the twice-daily ebb and flow of the tide. Even though estuaries can be bleak and often inaccessible places, they are, none the less, intriguing habitats. It is here that huge flocks of waders congregate to feed during the winter months, while eels and salmon make spectacular migratory journeys through estuarine waters. An estuary is not a cosy place for life, and yet, those plants and animals which have adapted to life in this habitat face very little competition from other species, and so are often present in very large numbers and high densities. This applies equally well to the worms and molluscs, as to the waders which feed on them.

As the sea water flows up the estuary, it meets and mixes with the fresh water downflow. Since fresh water is lighter than sea water, it tends to flow seawards out of the estuary above the incoming denser sea water. Mixing of the two types of water does take place where they meet, so that a middle layer of slightly salty brackish water forms. The amount of water carried down the river, and the state of the tide, both affect the rate and distance which the brackish water penetrates upstream. In a very dry summer, as in 1976, when the freshwater flow is greatly reduced, the brackish water will penetrate much further up an estuary than when a river is in full spate. Estuarine mudbanks develop from the sediment carried downstream by the river.

Lugworms, shore crabs, cockles and mussels are some of the more familiar seashore animals which are able to withstand the wide fluctuation of the salt content, or salinity, of sea water in estuaries. They can, therefore, be found some distance upstream from the open coast. There are also a few freshwater animals including three-spined sticklebacks, freshwater shrimps, and water lice, which are able to tolerate living in brackish water. There are, however, very few true brackish-water species which live in estuaries, but not in fresh water or the sea. Two of these are the tiny *Hydrobia* snail (18 and 19) and the opossum shrimp (30).

Estuaries not only provide important feeding-grounds for many birds but they are also valued sites for ports, oil-refineries, and power-stations. The concentration of industry increases the likelihood of pollution in estuaries. Many of the permanent animal inhabitants, however, are so well adapted to their fluctuating environmental conditions, that they can withstand a wide range of physiological stresses.

Figures in red beside a picture show the scale of reproduction.

2. ESTUARINE PART OF GILLAN CREEK, CORNWALL. From a high viewpoint beside the road, the expansive mud-flats of this Cornish valley can be seen at low tide. The steep rocky sides are typical of the drowned valleys of the area, caused by the rise in sea-level which followed the melting of the ice-caps after the last glaciation. In this estuary, gulls come to feed on the exposed mud which is rich in worms and shellfish. During low tide, the river continues to flow down the left side of the estuary. This site provides a sheltered anchorage for boats. It opens out south of the Helford River into Falmouth Bay at St Anthony, which is a popular place for cockle-gatherers.

3. SOLVA ESTUARY, DYFED. This attractive Welsh estuary is seen here in June when red valerian flowers on the wall beside the road. The sheltered creek leads into a fjord-like harbour, which is a popular boating centre. The Pembrokeshire Coast Path has to make a long detour round this indented harbour.

4. SALCOMBE ESTUARY, DEVON in August. Formed by the drowning of the coastline, this estuary extends far inland with many branching side creeks. Within these sheltered creeks, the rise and fall of the tide takes place so gradually that trees are able to grow right down to the water's edge without being damaged by salt spray. On this stretch, near Kingsbridge, there is an extensive growth of the green alga *Enteromorpha* on the mud-flats. Salcombe, which is the most southerly resort in Devon, has a mild climate. The surrounding land is extensively farmed, as can be seen by the field mosaic in the distance.

2

5
6

7

5 and 6. THE SEVERN BRIDGE. Like the open coast, twice a day estuaries are subjected to the rise (flow) and fall (ebb) of the tide. This pair of pictures illustrate how the tidal level changes the general appearance and mood of an estuary. The tidal range is particularly dramatic in the Severn Estuary where it can rise and fall more than 12 metres (40 feet) at Avonmouth. Tides are caused by the gravitational pulls of the sun and the moon on the sea water covering the earth. Even though the sun is much larger than the moon, it exerts a much smaller influence because it is so much further away from the earth. The rhythm of the tides is, therefore, governed chiefly by the lunar cycle. The side of the earth which is nearest to the moon will have a high tide, as will the part furthest from the moon. This huge tide 'bulge' or wave is constantly on the move as it passes round the earth. At the time of the new and also the full moon, when the sun's pull combines with the moon's pull, the high-ranging *spring* tides result. Spring tides occur at fortnightly intervals throughout the year, and have no connection with the spring season. The small-ranging *neap* tides alternate with the springs.

7. THE SEVERN BORE at Newnham. Bores form when the incoming flood tide passes through the narrowing banks of an estuary so that it piles up as a wall of water. This occurs only at the time of certain high-ranging spring tides. Before a bore is sighted the rush of the water can be heard. Seen from upstream, an advancing bore appears as a huge wave several feet high. For this reason, it is dangerous to stand too close to the water's edge when watching a bore. In October 1966, the Severn bore rose to 2·8 metres (9·25 feet) between Awre and Gloucester, as it advanced at a rate of 21 km/h (13 m.p.h.).

8. SPIRAL WRACK (*Fucus spiralis*) at Portishead. Seaweeds are able to grow in estuaries between the tides only where they can attach to rocks, piers, or breakwaters projecting above the mud. A seaweed is not anchored by roots, but by a flattened disc known as a 'holdfast'. Spiral wrack, which is related to bladder wrack, has no air bladders. The swollen ends visible here are the reproductive bodies.

9. EEL GRASS (*Zostera* sp.) with cockle. Growing both in muddy and muddy/sandy parts of some estuaries, eel grass has underground creeping rhizomes which can help to stabilise mudbanks. Eel grasses are the only flowering plants to grow in the sea in Britain. Certain areas, notably Chichester Harbour and Foulness, are frequented by over-wintering brent geese which feed on eel grass. Dried eel grass was once used to stuff mattresses.

8

9 ×0·3

10. TOWNSEND'S CORD-GRASS OR RICE-GRASS (*Spartina townsendii*) flowering. This species of grass is a natural hybrid which arose from an American species (*S. alternifolia*) crossing with a native one (*S. maritima*). First recorded in Southampton Water in 1870 it colonises bare mud higher up the estuarine shore than the eel grass. The cream flowering spikes appear in July and August.

11. TOWNSEND'S CORD-GRASS colonising exposed mudbank beside the Beaulieu River. This perennial plant grows vigorously, forming metre-high tussocks over tidal mud-flats, trapping silt between its roots and leaves. For this reason, it has been exploited commercially and introduced into estuaries both to help stabilise mudbanks and to reclaim land. Once it spreads over the mud it can transform it into grazing land within a few years.

10 ×1

11

12

13

12. PAGHAM HARBOUR, SUSSEX at low tide showing drainage channels. The raised banks have been colonised with glasswort (16) and sea purslane (17). Glasswort will tolerate being submerged for a short period each day; whereas sea purslane prefers well-drained banks. Salt-marsh areas develop at the upper end of sheltered estuaries or tidal creeks where there are no strong tidal currents. As the incoming tide creeps up over the flats it carries suspended mud and silt. These fine particles become deposited when the movement of water slows right down before it ebbs back down the shore. Plants which will survive a regular dousing by sea water are known as 'halophytes'. Such plants are relatively few in number, so that the variety of salt-marsh flowers is limited. None the less, the plants which thrive in salt marshes, such as sea thrift, sea lavender, and sea aster, can produce spectacular carpets of colour. In some places, salt marshes are referred to as 'saltings', especially the higher parts which are used for grazing.

13. PAGHAM, SUSSEX at low tide looking towards the coast. Water continues to flow down the creek throughout low tide. Marine snails, worms, and crabs, which become exposed on the mudbanks and mud-flats, are exploited by birds, such as the shelduck on the left of the picture. Extending down from the raised areas are broad banks of a green seaweed carpeting the mud surface.

14. ESTUARINE MUD. While it is quite safe to walk on the banks or on flat areas which have been colonised by cord-grass, it is unwise to venture far over bare mud. At best, the going is slow and tedious; at worst, it may mean abandoning a pair of gumboots! This footprint reveals that the mud has turned black below the surface. This is caused by certain bacteria producing hydrogen sulphide (smells of rotten eggs) which reacts with iron salts in the mud to form a black substance known as 'ferric sulphide'. Black mud (or sand) is, therefore, an indication of anaerobic conditions.

14

15 ×0·2
16 ×0·6

17 ×3

15. SEA PURSLANE (*Halimione portulacoides*). The margins of drainage channels are often bordered by a wide band of greyish-green sea purslane. This perennial plant can form extensive swards over the high well-drained parts of the marsh. Sea purslane is most common on the east coast and is very local on the west coast. In this picture some green seaweed deposited by the tide can be seen at the edge of the channel.

16. MARSH SAMPHIRE OR GLASSWORT (*Salicornia europaea*). This succulent annual plant may turn a spectacular red colour with the autumn frosts. Marsh samphire colonises bare mud where conditions are fairly stable. Minute yellow flowers are produced in the axils of the fleshy stems. The name 'glasswort' arose from the soda ash, obtained from burning the plants, which was used for glass-making. This is perhaps an unlikely edible plant, but it can be eaten both fresh in salads or boiled like asparagus. June and July are the best months for gathering plants to be eaten as a salad vegetable; whereas the end of the summer will not be too late for collecting plants for boiling. Usually growing to about 15 centimetres (6 inches) high, a giant 2 metres (6½ feet) high specimen grew up in a Norfolk creek after the 1953 floods had washed down nutrients from surrounding farmland.

17. SEA PURSLANE (*Halimione portulacoides*) in flower. Although far from showy, the yellow flowers on short spikes are easier to spot than the tiny glasswort flowers. Sea purslane leaves can also be eaten in salads. A coating of white scales gives the leaves their silvery-white colour.

18. HYDROBIA SNAILS (*Hydrobia ulvae*) on glasswort. These tiny 6 millimetre (0·25 inch) long snails are one of the most abundant estuarine molluscs. Densities of up to 42,000 per square metre have been recorded from the Clyde Estuary. They can be found exposed on the mud (26) as well as on plants as in this picture. They adhere to vegetation as the film of water round their shell dries out. They can, therefore, easily be knocked off the glasswort or any other plant. The average salinity of the sea is 35‰, and *Hydrobia* will tolerate a range from 10–33‰ like many other estuarine inhabitants.

19. HYDROBIA SNAILS CRAWLING UP GLASS. Each snail crawls on its muscular foot, with its tentacles outstretched. On all the shells there is a slime coating mixed with silt. Although *Hydrobia* are often associated with the green seaweed, sea lettuce (*Ulva lactuca*), the snails browse on a variety of seaweeds as well as feeding on microscopic plants known as 'diatoms' and also detritus on the mud surface. In turn, the huge concentrations of *Hydrobia* provide an important food source for short-billed wading birds, as well as shelduck and pintail. The birds feed most actively as the tide ebbs down the shore when the snails themselves become exposed as they crawl and feed over the mud surface. The snails gradually burrow into the mud, only to reappear when the tide flows in again. They then bob up to the surface, where they are kept afloat by a mucus raft which also traps diatoms for food. The *Hydrobia* eggs, which are laid in gelatinous capsules, hatch into planktonic larvae which are carried in the surface waters until the last larval stage changes into a small snail.

20 ×0·3
21 ×0·2

20. SHORE CRAB (*Carcinus maenas*) among eel grass. A common inhabitant of rocky shores, this crab also lives in salt marshes and estuaries, where it will tolerate a salinity as low as 6‰. The shell colour is often green as shown here, but it may also be brown or orange. Shore crabs feed both as predators, on live prey, and as scavengers on rotting carcasses.

21. MOULTED SHELL OF SHORE CRAB. The hard outer skeleton prevents crabs from gradually increasing their size. Instead, they periodically shed their shell by crawling out backwards. Freshly moulted crabs have soft shells which make them easy prey to predators for a few days. This picture shows a discarded crab shell, or carapace, lying in a salt marsh. Sometimes large numbers are washed up along the high tide mark. Most crabs have one pair of pincers and four pairs of walking legs; the missing lower leg on the left of this photograph is tucked away out of sight.

22. DRIED SALT-PAN. Small muddy depressions with no vegetation are features of salt marshes. These salt-pans may be formed in one of two ways. When plants begin to colonise the salt marsh, they may spread unevenly, so that raised hummocks join up, thereby enclosing an area of bare mud which cannot be drained by the ebbing tide. Larger, elongated pans are formed by the blockage of a creek. During spring tides, when the tide washes high up the marsh, the pans become filled with water. During neap tides, however, the water evaporates, leaving a very salty solution. It may even dry up completely, like this pan which was photographed in Pembrokeshire, in June.

23 ×0·3
24 ×3

23. EDIBLE WINKLES (*Littorina littorea*). This discarded shoe is appreciated by the winkles, which can browse on the pinkish algal growths. Edible winkles are both the largest and commonest of the shore winkles. When the snail withdraws into its shell, it seals the opening with a horny plate known as an 'operculum'. In the spring, female winkles lay their egg capsules at night as the tide rises up the shore. Each capsule contains 1–5 eggs.

24. EDIBLE WINKLE crawling inside an aquarium. Notice the characteristic cross-bands on the pair of tentacles and the line running down the centre of the foot. Winkles move by a double series of muscular waves (one on each half of the foot) passing from the front end backwards.

25. COCKLE SHELLS (*Cerastoderma* (=*Cardium*) *edule*). Large commercial cockle-beds are sited in estuaries, notably the Thames and the Wash. At low tide, the cockles burrow just below the surface, moving up to feed when covered by the tide. They feed by drawing water in through one siphon and filtering off the food particles, before ejecting the waste current through a second siphon. Under favourable conditions, as many as 10,000 cockles per square metre can occur.

26. LIVE COCKLES WITH HYDROBIA SNAILS. Still covered with a layer of water, one of these cockles has not yet closed its shells tightly together. Oystercatchers feed on cockles, and this fact, coupled with the decline of the cockle harvest in the Burrey Inlet in South Wales, led to the controversial culling of many oystercatchers in 1974.

25 ×1
26 ×0·3

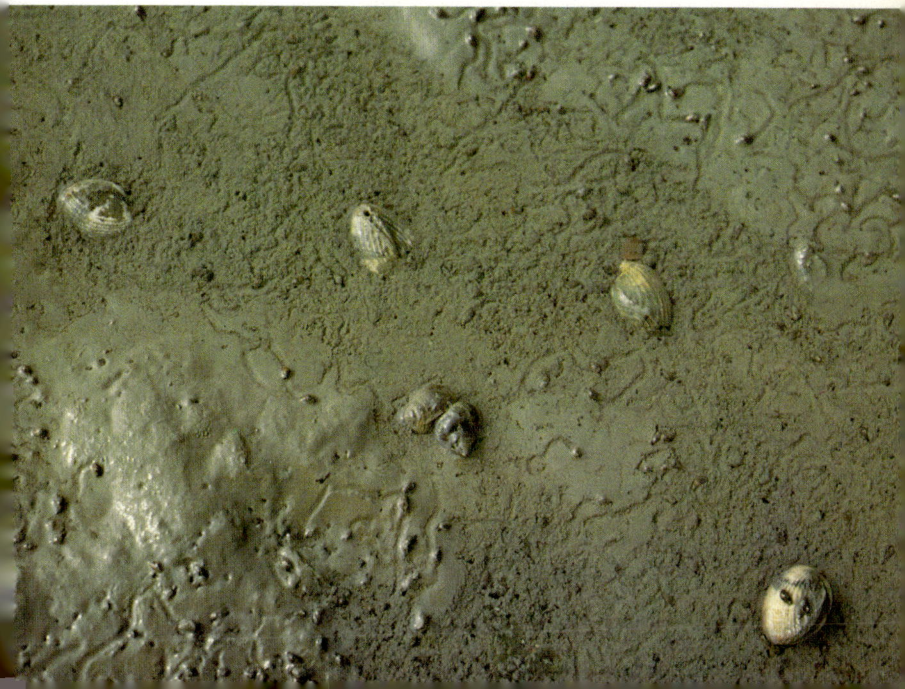

27 ×0·3

28 ×0·4

27. LUGWORM (*Arenicola marina*) casts and depressions. Like earthworm casts on a lawn, these lugworm casts pinpoint the location and density of the worms living below the beach surface. Lugworms live in a burrow, periodically moving up towards the surface to form the cast at the tail end. The conical depressions are formed by the sand sinking downwards as the lugworm swallows sand for as much as eight hours a day. Most of what is swallowed gets pushed out through the gut as the familiar cast, and only a minute fraction is absorbed as food. Lugworms are a popular bait for sea fishermen, who can often be seen digging for worms at low tide.

28. MUSSELS (*Mytilus edulis*). Where there is a constant supply of microscopic organic debris and planktonic organisms, such as in sheltered estuaries, mussels form extensive beds. These mussels, which are exposed to the air, have closed their shells to prevent their insides from drying out. As soon as they are covered by the tide, the shells open so the mussels can begin to feed. Water and food particles are drawn inside by a mass of tiny hairs (cilia) on the orange gills, beating to create a current. The food is trapped in a mucus film on the gills and the waste rejected in a stream of water. Mussels lower down the shore are able to grow faster, because they are covered for longer periods and can, therefore, feed for a long time each day. Mussels spawn in late spring. The fertilised egg develops into a 'veliger' larva which lives a planktonic existence for some weeks. It then changes into a 'periveliger', which settles on to a solid bottom, such as rocks or buoys, often where other mussels are already present.

29. LUGWORM (*Arenicola marina*). The three distinct parts of the worm: the head, trunk, and tail are shown here. Arising from the trunk region are thirteen pairs of red feathery gills. The lugworm is starting to burrow – head end first – into the sand.

29 ×1

30 × 5

30. OPOSSUM SHRIMPS (*Neomysis integer*). Also known as 'mysids', these crustaceans are not true shrimps. *Neomysis* is particularly abundant in estuaries. It is an important link in estuarine food webs, since it feeds on a variety of food and is itself eaten by many fish. Since opossum shrimps spend most of their time swimming actively in mid-water, instead of crawling on the bottom, they are termed 'planktonic'. They often congregate in dense populations at the edge of the incoming tide. They feed on other planktonic organisms, including tiny plants (phytoplankton) and animals (zooplankton) by filtering them from the water. They also feed by scavenging on larger crustaceans. *Neomysis* will not survive in polluted estuarine water. Female opossum shrimps carry their young in a brood pouch which can be seen extending behind the legs in the central animal. There may be as many as six broods produced during the year. In the winter, however, the broods are smaller (ten to twenty-five) than in the summer (twenty to fifty). Opossum shrimps are sold by aquarists as live food for marine fish.

31. MASSED SLUDGE WORMS (*Tubifex* sp.). These worms are also used as live food by aquarists. They will tolerate very low salinities and so are used as food for freshwater fish. Related to earthworms, sludge worms flourish in muddy estuarine banks even in quite polluted places. They also abound in the filter-beds of sewage farms. Because *Tubifex* can tolerate not only polluted conditions, but also low oxygen levels, populations of these worms are able to build up quickly without being kept in check by either competitors or predators. At times, the mudbanks of the Thames Estuary can appear distinctly red from masses of these worms. Each worm builds a chimney-shaped tube from the mud. It lives head-down in the tube, with the tail emerging. The tail beats to and fro creating a water current towards the head. The worms are collected by digging out the mud and placing it on to strips of hessian above buckets of clean water. The worms separate from the mud by wriggling through the hessian and dropping into the water below.

31 × 4

32. ORAL DISC OF SEA LAMPREY (*Petromyzon marinus*). Lampreys are primitive jawless fish which have a circular sucking mouth. The body, which is eel-like, has no paired pelvic or pectoral fins. The gills open directly to the outside by seven gill slits. The sea lamprey, which is the largest of the three British species, has a much more formidable array of teeth than either the river or the brook lamprey. It is a parasite which attacks several kinds of fish – including salmon – by clamping on to the skin with its sucking disc. The flesh of the host is then rasped away by the toothed tongue. The lamprey produces a substance which prevents the victim's blood from clotting. When small fish are attacked by a sea lamprey, the result can be fatal. Sea lampreys move from the sea up estuaries to spawn in fresh water in May and June, making a nest by removing stones with their mouth disc. The adults die after they have spawned. The larvae, when they hatch, live buried in the bottom. They are blind and toothless, so they do not lead a parasitic life; instead they feed on organic matter.

33. SALMON PUTCHEON WEIR in the Severn Estuary. Salmon make spectacular migrations through estuaries. The adults move upstream to spawn and the immature fish move downstream out to sea, where they live for several years before the mature fish make the return journey. Many of the salmon from Britain travel westwards up to the waters around west Greenland to feed. Salmon are caught in a variety of ways, In rivers, the fly fisherman has to select the correct artificial fly which will successfully mimic the natural fly on which the fish are feeding. In estuaries, Seine nets are usually used for catching salmon. In the Severn Estuary, however, putcheon weirs are still used. These consist of a series of wickerwork funnel-shaped baskets tied to a weir of stakes, which run out at right angles to the estuary. The mouth of each putcheon or 'putt' faces upstream, so that the salmon swims down the narrow-shaped funnel. The fishermen collect the salmon from their 'putts' at low water.

34. SALMON (*Salmo salar*) caught in a Seine net in the Hampshire Avon Estuary.

32 ×6·5

33

34

35 ×2·5
36 ×0·5

35. SALMON (*Salmo salar*) alevins. Both salmon and trout lay large eggs, from which the alevin stage hatches with a large yolk sac. These young salmon were photographed at a hatchery four days after hatching. The yolk sac is gradually absorbed during the first month, by which time the salmon starts to feed on live food and is known as a 'fingerling'. Under natural conditions, salmon spawn in November and December. The female or hen fish cuts out the hollow, or redd, in gravel, by flexing her body rapidly over the bottom so that the silt and smaller pieces of gravel get swept downstream. When the redd is completed, the male or cock fish swims up close beside the hen fish. In this way he can fertilise her orange eggs, as they are laid, with his white milt. After spawning, the spent fish, which are quite thin, are known as 'kelts'. Most kelts die, only a few survive to go downstream to sea and to return upstream to spawn for a second time.

36. SALMON (*Salmo salar*) smolt. This stage, which develops from the parr illustrated on the opposite page, has silvery sides. Smolts move downriver out to the sea via the estuary in May and June.

37. SALMON (*Salmo salar*) parr. The fingerlings grow for about a year until they become parr, which have a series of dark blotches down each side of the body.

38. SALMON (*Salmo salar*) scales. The silvery-blue colour of these scales is typical of a fresh-run salmon which has recently moved in to a river from the sea. A further indication of a salmon's recent arrival from the sea is the presence of external copepod parasites called fish lice.

37 ×1

38 ×3

39 ×2
40 ×1·5

39. EEL LEPTOCEPHALUS LARVA (*Anguilla anguilla*). This stage in the eel's life-history first appears in the Atlantic Ocean in the Sargasso Sea area, where the adult eels congregate to breed. When the flattened, transparent leaf-like larvae were first discovered in 1856, they looked nothing like eels as we know them. It was hardly surprising, therefore, that they were given the new scientific name of *Leptocephalus*. It takes two and a half to three years for the larvae to drift across to Europe.

40. EEL ELVERS (*Anguilla anguilla*). A few months before the larvae reach our river-mouths, they change into round-bodied elvers or glass eels. At first, the bodies are completely transparent, so that the heart, gills, gut, and backbone can be seen clearly.

41. EELS (*Anguilla anguilla*) caught in eel-trap. As the elvers move upstream, their bodies gradually darken. Eels live in fresh water for several years before they move downstream on their spawning migration towards the end of the summer. Their yellow bellies then turn silver and their eyes enlarge. At the same time as the reproductive organs enlarge, the gut begins to break down so that the eels cease to feed.

42. ELVER FISHERMAN WITH NET. In the Severn Estuary, elvers are caught in a special net. Nylon mesh is stretched over the wooden framework. The elvers are caught shortly after the tide has begun to ebb, by holding the mouth of the net facing the sea. The young eels swim into the net as they move upstream against the current. Live elvers are exported to stock eel ponds.

41 ×0·2

42

43
44

43. BRENT GEESE (*Branta bernicla*). These geese move south to over-winter in some of our south-east estuaries in particular. Their favourite food is eel grass (9). There are two distinct populations of brent geese, the light-bellied form and the dark-bellied form shown above.

44. MALLARDS (*Anas platyrhynchos*). These duck were feeding early in the morning in a muddy estuarine creek. While mallards are widespread in almost any type of habitat, they often move towards the coasts during winter.

45. MUTE SWANS (*Cygnus olor*). Broad-webbed feet prevent the swans from sinking too far into the estuarine mud-flats. Like mallards, mute swans are equally at home on inland waters as in estuaries.

46 ×0·5

46. FLOUNDER (*Platichthys flesus*). This is the only flat fish in Britain which can move up through estuaries to live in fresh water for a short time. The upper surface has a mottled brownish coloration which helps the fish to blend in with the bottom. Additional disguise is gained by the flounder flipping sand up over its body. The underside, which is seen only when the flounder swims up off the bottom, is white. Flounders return to the sea to breed in the spring in water 27–54 metres (88–170 feet) deep. The small eggs, only about 1 millimetre (0·04 inch) in diameter, float up to the surface. As the embryo develops, the eggs slowly sink until after a week they hatch into symmetrical larvae with an eye on each side of the head. On metamorphosis, the left eye moves round so that both eyes lie on the upper, right side when the flounder rests on its left side on the bottom. Young flounders feed on small crustaceans, while larger fish feed on molluscs, including mussels, cockles, and Baltic tellins, as well as worms and crustaceans. These flat fish can be a serious pest of commercial cockle-beds. Although flounders are not a popular food fish in Britain, they are an important commercial fish in both the Baltic and off Denmark.